Radu Vasilache, Maria Toader

GRIN Verlag

Monitoring of environmental factors in a radioactive polluted area in Romania

GRIN Verlag

Bibliografische Information der Deutschen Nationalbibliothek:

Die Deutsche Bibliothek verzeichnet diese Publikation in der Deutschen National-
bibliografie; detaillierte bibliografische Daten sind im Internet über http://dnb.d-
nb.de/ abrufbar.

Imprint:

Copyright © 1998 GRIN Verlag GmbH
Druck und Bindung: Books on Demand GmbH, Norderstedt Germany
ISBN: 978-3-656-69565-3

This book at GRIN:

http://www.grin.com/en/e-book/276308/monitoring-of-environmental-factors-in-
a-radioactive-polluted-area-in-romania

Maria Toader, Chief Scientific Investigator
Radu A. Vasilache, Senior Physicist

MONITORING OF ENVIRONMENTAL FACTORS IN A RADIOACTIVE POLLUTED AREA IN ROMANIA

NATO Project CGR 972725

Institute of Public Health Bucharest
1998

Table of contents

1. INTRODUCTION

Within the last years, a great deal of interest was focused on studies of environmental pollution, and particularly on the pollution levels near fertilisers plants, with the possible the consequences for the health of the population living in those areas. The preparation of the phosphate fertilisers, has the unfortunate potential for environmental pollution. This is due to the radioactive content of the phosphatic rocks used as prime matter for fertilisers.

The radioactivity present in phosphate rocks is mainly caused by the daughter products of U-238. In fact, natural uranium can substitute for calcium in the phosphate rock structure and, over a period of time, accumulate in the phosphate reserves.

The behaviour of U-238 and Ra-226 (the principal contributors of radioactivity in phosphate rock) could be interpreted as follows: Ra-226 is precipitated with the solid phase (phosphogypsum) as insoluble radium sulphate, while U-238 remains in the liquid phase, as a uranil complex. The presence of higher levels of radioactivity in superphosphate, which contains the radium associated with the original phosphate rock and precipitated as phosphogypsum has become a cause for environmental concern.

The products and by-products of the phosphate industry contain various fractions of the uranium, radium and thorium originally present in the rock, depending on the manufacturing process.

The present study is aimed towards identifying the source of pollution and the evaluation of the level of radioactive pollution of the environmental factors and the impact on the population in the area of the phosphatic fertilisers plant ROMFOSFOCHIM - Valea Calugareasca.

The industrial activity in the Valea Calugareasca area started in 1902 by the creation of the Romanian Society of Chemical Products. The factory initially produced sulphuric acid, then started the production by-products of the sulphuric acid and fertilisers. After 1950, the factory diversified the production, producing in 1990 more than 10 different products, some of them vital for the Romanian economy. In the summer of 1997 the company ROMFOSFOCHIM was closed.

To produce sulphuric and phosphoric acid, ROMFOSFOCHIM Valea Calugareasca imported from Morocco millions of tons of phosphatic rock (justified by economic reasons - due to the low cost) with a high radioactive content.

In the manufacturing process the main waste resulted is the phosphogypsum, which is a radioactive waste deposited by ROMFOSFOCHIM in open sites spread over tenths of kilometres around the factory. Other wastes resulted from the manufacturing process of the sulphuric acid are quite large quantities of pyritic ashes, a blend of iron oxides, fluoride and arsenium salts and heavy metals oxides.

With a total disregard of the dangers generated by the presence of the toxic and radioactive wastes, ROMFOSFOCHIM managed to spread on hundreds of hectares mountains of pyritic ashes, more than 50 meters high and containing millions of tons of toxic wastes. Both the ash mountains and the phosphogypsum repositories (occupying more than 50 ha) were left without any surveillance and uncovered. As the radioactive phosphogypsum looks like normal cement and is an excellent ligand, the local population, unaware of the dangers, used it for repairing their houses, in stead of other building materials. The obvious result were houses with a higher radon content, in which they continue to live.

In August 1997 the activity at ROMFOSFOCHIM was stopped and the factory closed down its gates. The tenths of hectares covered with toxic and radioactive wastes and hundreds of polluted hectares remained. A visitor can find now a real moon-like view around the factory. The vegetation is destroyed, the soil is calcinated, there are mountains of pyritic ashes spread around on tenths of kilometres by the wind, repositories with radioactive wastes, all of them destroying slowly the life in the area.

2. MATERIALS AND METHOD

The samples were prelevated from the surroundings of a phosphate plant located near Valea Calugareasca - see the map.

2.1. Types of samples

2.1.1. Environmental samples

- the drinking water samples were prelevated directly from the private sources used by the population;
- water waste samples were prelevated from two different sites, up and down the river by respect to the plant;
- the   soil samples were prelevated from areas of 500 cm^2 and from the widths 0-5 cm ;
- the wild vegetation and cultivated plants were prelevated from the entire area of study, taking into account the geographical placement of the factory and the specifics of the agriculture in the area ;
- fallout was prelevated for periods of one month, from surfaces of 1 m^2;
- the aliments are specific to the area: milk, cheese, meat, bread, fruits and vegetables.

2.1.2 Biological samples

- embryos and foetuses prelevated from women having abortion, and living in the area of interest

2.2. Prelevation sites

- from the new phosphogypsum repository site, near ROMFOSFOCHIM plant;
- from the old phosphogypsum repository site, near Darvari village;
- from the villages Pantazi, Berceni and Costegi, situated near the two phosphogypsum repositories;

- background samples for environmental factors were prelevated from Valenii de Munte village;
- the background samples for the radioactive content of foodstuff were prelevated from Dracea village (Teleorman county), having roughly the same population and dietary habits as in the area of interest and with no radioactive pollution sources nearby).

2.3. Radioactive pollution indicators

Based on a thorough analysis of:
- the polluting radionuclides (from the point of view of the toxicity and the biological and physical half-time),
- the most probable paths for pollution ,
- the geographical placement of the factory,
- the specifics of the agriculture in the area,
- the particularities of the population diet in the area,

we have chosen as indicators for radioactive pollution in the area the following radionuclides: Ra-226 and Po-210.

We should mention a few things on the radiotoxicity of these elements:

- Po-210 has been considered to be five times more toxic than Ra-226 and as hazardous as plutonium by some authors (Fink 1950, Blanchard 1967, McDonald et al. 1986). Stanard (1988) concluded: "Polonium is obviously a much stronger candidate than plutonium for the title of <<most toxic element known to man>> in terms of acute lethality on either a weight basis or an activity basis." Po-210 is also mobile in the body and has higher dose rate than either Ra-226 or Pu-239 because of its short half-life (138 d).

- Ra-226 is another radionuclide of high interest, due to its radiotoxicity and very long half-time. It has been demonstrated that, in humans who were exposed to radium as adults, radium retention 30-60 years later depended on the quantity of radium deposited within their bodies. Therefore it should be pointed out that Ra-226 intake leads to a dose commitment for a long period of time. A careful reading of ICRP Publication 20 (1973) suggests that the retention might as well be related to the level of radium deposited within the body, which in turn is related to the intake, which is related to the level of environmental pollution in the area of residence.

2.4. Radiochemical separation of Po-210 and Ra-226.

2.4.1   Radiochemical separation of Po-210.

Polonium is quantitatively deposited on a nickel disc from strong HCl solution . This is a very specific separation and therefore can be carried out while many other radionuclides are present in the sample.

2.4.2 Radiochemical separation of Ra-226

2.4.2.1   Radium-226 in Water and Urine

Radium is isolated from most other elements by coprecipitation with $BaSO_4$. Further purification is obtained by the removal of silica with HF and reprecipitation of the sulphate. The sulphate precipitate is dissolved in alkaline EDTA to prepare the emanating solution.

2.4.2.2   Radium-226 in the Soil, Vegetation , Food and Milk

A variety of matrices may be prepared for ^{222}Rn emanation measurement. For those matrices requiring a Na carbonate fusion, the alkali metals are leached from the ground fused material with hot water. The insoluble alkaline earth carbonates are dissolved in HNO_3. A phosphate collection precipitate is used on ashed samples which can be dissolved directly in mineral acid. This precipitate is then dissolved in HNO_3. Successive fuming HNO_3 separations remove the Ca and most other interfering ions. Traces of some fission products are scavenged with Y hydroxide. Finally, Ra is coprecipitated with Ba chromate, then dissolved in HCl and H_2O. $RaBaSO_4$ is precipitated, filtered and radiometric measurement

2.5 Radiometric measurement of Po-210 and Ra-226

The radiometric measurement of Ra-226 and Po-210 is performed with an alpha counting system, having an alpha probe with excellent detection characteristics and low background. Radium-226 in solution were determining by de-emanating its Rn-222 progeny into an scintillation cell for measurement.

The results are expressed in Bq/l for liquid samples and Bq/kg for solid samples (Bq/kg DW and Bq/kg GW).

2.6 Annual effective dose

In order to determine the internal irradiation in Valea Calugareasc and the neighbouring area we have determined the intake of natural radionuclides by ingestion of contaminated foodstuff (with Ra-226 and Po-210).

The choice of this method for estimating effective doses through use of measurements of ingestion of radionuclides is justified by the higher accuracy compared with use of body contents of radionuclides (UNSCEAR 1993). Furthermore, the estimation of internal exposures after the determination of radionuclides in the diet an food samples is easier, more rapid, and more useful in an emergency situation than the procedures using the analysis of human tissues and organs.

The annual effective dose is calculated with the ingestion dose coefficients (Sv/Bq) reported by the International Commission on Radiological Protection (ICRP 68, 1994).

In order to determine the internal irradiation in Valea Calugareasca and the neighbouring area we have determined the intake of natural radionuclides by ingestion of contaminated foodstuff (with Ra-226 and Po-210)and to evaluate the risks health.

3. RESULTS

The results obtained are presented in Table 1 separated by types of samples and location prelevation :

TABLE 1 The Ra-226 and Po-210 content of enviropnmental factors in the area near ROMFOSFOCHIM Valea Calugareasca

Nr crt	Preleva-tion sites	Type of samples	Nr. of sam-ples	Ra-226 mBq/l, kg (GW)	Po-210 mBq/l, kg (GW)
1.	Near ROMFOS-FOCHIM plant	drinking water	14	11÷60	6÷32
		waste water	10	515÷980	364÷704
		fallout[1]	10	114÷1556	32÷930
		soil	16	88311÷121650	56432÷70114
		wild vegetation	16	530÷2511	302÷1820
		cultivated plants	10	104÷582	43÷270
2.	Darvari	drinking water	15	10÷73	5÷31
		fallout[2]	10	82÷1420	30÷420
		soil	18	73510÷132311	40618÷82190
		wild vegetation	13	553÷2815	269÷1158
		cultivated plants	15	98÷435	56÷142
3.	Village Pantazi	drinking water	20	10÷81	6÷59
		fallout[†]	10	56÷1139	33÷520
		soil	15	78431÷114568	37241÷71829
		wild vegetation	18	210÷2714	106÷1315
		cultivated plants	14	48÷520	28÷135
4	Village Berceni	drinking water	13	8÷56	5÷42
		fallout[†]	10	46÷1060	28÷248
		soil	11	68513÷87919	40512÷56172
		wild vegetation	19	188÷2130	88÷1011
		cultivated plants	21	50÷114	34÷88
5	Village Costegi	drinking water	12	8÷50	5÷42
		fallout[†]	10	51÷930	28÷246
		soil	10	63114÷92529	37785÷42511

[1] In Bq/m² per month
[2] In Bq/m² per month

2.		wild vegetation	15	160÷1810	56÷657
		cultivated plants	11	42÷153	33÷105
6	Village Valenii de Munte (back-ground samples)	drinking water	10	4÷8	<3
		fallout[†]	10	11÷32	3÷10
		soil	15	26200÷30140	11040÷13050
		wild vegetation	13	83÷148	19÷46
		cultivated plants	18	27÷83	13÷38

From the analysis of the data in table 1, we can conclude the following comparing to the values from the control area, the area near ROMFOSFOCHIM is subjected to radioactive pollution. The mean values from the monitored area are 2÷3 times higher than the values from the control area.

The large range of values is due to the meteorological factors affecting the unprotected waste dunes. The wind and the rain are continuously redistributing the waste, therefore the values in different points depend upon the season and the weather.

All the prelevation points from presented in Table 1 are included in the area of influence of the two waste repositories, having the mean values of the radioactive content presented in Table 2.

Table 2   The Ra-226 and Po-210 content of the soil and vegetation from the two waste repositories

Nr. crt.	Prelevation sites	Type of samples	Depth	Ra-226 Bq/kg (GW)	Po-210 Bq/kg (GW)
1.	New repository site	Soil	0-2 cm	134±33	41±10
			2-5 cm	185±41	52±15
			5-10 cm	111±34	36±8
		Wild vegetation		8±3	7±3
2.	Old repository site	Soil	0-2 cm	53±12	31±5
			2-5 cm	61±15	33±8

		5-10 cm	83±15	56±10
	Wild vegetation		5±1	4±1

As it can be seen the wild vegetation growing on the waste dunes has a very high radioactive content. By analysing separately the radioactive content of the roots and stalk, it can be seen that Ra-226 is distributed 60% in roots and 40% in the stalk, while the Po-210 is roughly 75% in the roots and 25% in the stalk.

The distribution through the plant is also influenced by the age of the dune. In the new repository site, the roots have a higher content, going up to 80% Ra-226 and Po-210 85% in roots. This can be explained by the depth distribution of the radionuclides (the surface layer of the soil has a higher contamination).

As the environmental factors indicate a radioactive pollution of the area, a high interest is presented by the internal contamination of the population living in the area. Therefore, the study was extended to the evaluation of contamination of the foodstuff, which, by intake, gives birth to the internal exposure of the population. To calculate the intake we had to identify the specifics of the diet in the area, to identify the most used foodstuff in the area. Based on the results of this study we have analysed the most frequently used aliments in the area and we have chosen a control area with a population having the same dietary habits for the adults and children.

The results of the determinations are presented in the tables below, separately for adults and children of different age groups: 1-3 years, 4÷6 years, 7÷9 years, 10÷12 years and youngsters aged of 13÷15 years and 16÷19 years.

Table 3 shows these results for the adults living in the area of interest, compared to the values for the intake in the adults living in the control area, while tables 4÷10 present the same comparison for chidren of different age groups.

Table 3: The amount of Ra-226 and Po-210 ingested by the adult population near the plant from Valea Calugareasca compared to a control area

Crt. no.	Food	Annual ingestion (kg)	%	Radioactive intake (mBq)			
				Ra-226		Po-210	
				Plant area	Control	Plant area	Control
1	Milk:	240		2937	2145	1881	1373
	cow		80	1651	1363	998	787
	sheep		20	1286	782	883	586
2.	Meat:	50		3211	1394	1559	1020
	beef		10	275	165	191	106
	pork		50	1005	525	515	370
	mutton		20	1228	453	565	382
	chicken		20	703	251	288	162
3.	Cereals:	220		13965	6479	6997	3282
	flour		75	12012	5610	5973	2706
	corn flour		25	1953	869	1012	576
4.	Eggs	20		2120	1520	684	410
5.	Vegetables	130		5097	1759	3379	1625
	carrots		55	3225	1008	2130	1015
	tomatoes		15	708	238	628	211
	cabbage		10	238	126	166	104
	root veg.		15	523	192	283	168
	leafy veg.		5	403	195	272	127
6.	Fruits:	80		3519	1112	2100	1083
	apples		40	2022	547	1069	582
	plums		10	340	74	130	82
	grapes		30	713	300	636	256
	quince		5	179	41	79	36
	cherries		15	265	150	186	117
	Total:			30849	14409	16600	8793

Table 4: Annual amount of Ra-226 and Po-210 ingested by 1-3 years old children, living in Valea Calugareasca area compared to those living in a control area

Crt. no.	Food type	Annual ingestion	Radioactive intake (mBq)			
			Ra-226		Po-210	
			Plant area	Control	Plant area	Control
1.	Milk	255	3120	2279	1998	1459
2.	Meat	15	963	418	468	306
3.	Cereals	36	2285	1060	1145	57
4.	Eggs	9	954	684	308	184
5.	Potatoes	30	1612	504	1015	508
6.	Vegetables	52	1390	647	1162	531
7.	Fruits	34	1496	473	892	460
		Total	11820	6065	6988	3985

Table 5 Annual amount of Ra-226 and Po-210 ingested by 4-6 years old children, living in Valea Calugareasca area compared to those living in a control area

Crt. no.	Food type	Annual ingestion	Radioactive intake (mBq)			
			Ra-226		Po-210	
			Plant area	Control	Plant area	Control
1.	Milk	290	3549	2543	2274	1659
2.	Meat	18	1156	502	561	367
3.	Cereals	58	3682	1708	1844	865
4.	Eggs	14	1484	1064	479	287
5.	Potatoes	50	2303	720	1450	725
6.	Vegetables	60	1872	238	628	211
7.	Fruits	50	2199	695	1312	677
	Total		16245	7520	8548	4791

Table 6 Annual amount of Ra-226 and Po-210 ingested by 7-9 years old children, living in Valea Calugareasca area compared to those living in a control area

Crt. no.	Food type	Annual ingestion	Radioactive intake (mBq)			
			Ra-226		Po-210	
			Plant area	Control	Plant area	Control
1.	Milk	290	3550	2593	2274	1659
2.	Meat	20	1284	558	624	408
3.	Cereals	68	4317	2003	2162	1015
4.	Eggs	18	1908	1368	615	369
5.	Potatoes	65	1612	504	1015	508
6.	Vegetables	61	1903	763	1371	620
7.	Fruits	58	2551	806	1522	785
		Total	17125	8595	9583	5364

Table 7 Annual amount of Ra-226 and Po-210 ingested by 10-12 years old children, living in Valea Calugareasca area compared to those living in a control area

Crt. no.	Food type	Annual ingestion	Radioactive intake (mBq)			
			Ra-226		Po-210	
			Plant area	Control	Plant area	Control
1.	Milk	290	3549	2593	2274	1659
2.	Meat	26	1670	725	811	530
3.	Cereals	98	6221	2886	3116	1462
4.	Eggs	18	1908	1368	616	369
5.	Potatoes	68	3133	979	1972	986
6.	Vegetables	68	2126	851	1529	651
7.	Fruits	60	2639	834	1575	812
	Total		21246	10236	11893	6469

Table 8 Annual amount of Ra-226 and Po-210 ingested by youngsters aged of 13-15 years, living in Valea Calugareasca area compared to those living in a control area

Crt. no.	Food type	Annual ingestion	Radioactive intake (mBq)			
			Ra-226		Po-210	
			Plant area	Control	Plant area	Control
1.	Milk	290	3549	2593	2274	1659
2.	Meat	31	1991	864	967	632
3.	Cereals	125	7935	3681	3975	1865
4.	Eggs	18	1908	1368	616	369
5.	Potatoes	73	3363	1051	2117	1073
6.	Vegetables	74	2309	926	1663	645
7.	Fruits	67	2942	931	1759	907
		Total	23997	11414	13371	7150

Table 9 Annual amount of Ra-226 and Po-210 ingested by youths aged of 16-19 years, living in Valea Calugareasca area compared to those living in a control area

Crt. no.	Food type	Annual ingestion	Radioactive intake (mBq)			
			Ra-226		Po-210	
			Plant area	Control	Plant area	Control
1.	Milk	290	3549	2593	2274	1659
2.	Meat	35	2248	976	1091	714
3.	Cereals	140	8887	4123	4452	2089
4.	Eggs	18	1908	1368	616	369
5.	Potatoes	70	3225	1008	2030	1015
6.	Vegetables	60	1872	751	1149	610
7.	Fruits	70	3073	973	1837	948
		Total	24762	11792	13449	7404

It can be seen that the cereals have an important contribution to the internal conatmination (about 30%), both by their high radioactive content and by their importance in the diet (which is specific to Romania). In the order of their contribution to the contamination, the cereals are followed by the milk and potatoes.

From the tables above, we can make a comparison between the results for the different ages. Table 10 shows this comparison between the daily intake in the adults and the intake in children of different ages.

Table 10 Daily intake of Ra-226 and Po-210 for subjects of different ages living in Valea Calugareasca area compared to those living in a control area

Crt. no.	Group composition	Age group	Radioactive intake (mBq/day)			
			Ra-226		Po-210	
			Plant area	Control	Plant area	Control
1.	Children	1-3	32.4	16.6	19.2	10.9
2.		4-6	44.5	20.6	23.4	13.1
3.		7-9	44.9	23.6	26.3	14.7
4.		10-12	58.2	28.0	32.6	17.7
5.	Youngsters	13-15	65.7	31.3	36.6	19.6
6.		16-19	67.8	32.3	36.9	20.3
7.	Adults		84.5	39.5	45.5	24.1

The mere daily intake is not very significant when considering the effects on the public health, therefore we have calculated the internal doses due to the ingestion of the two radionuclides, using the dose factors from ICRP 67.

These doses are presented, for all age groups, in table 11 below. As it can be seen from tables 10 and 11, although the quantity ingested is higher for Ra-226 than for Po-210, the doses due to Po-210 are significantly higher than those for Ra-226.

Also, the effective dose is, as expected, a function of age. The dose distribution as a function of age also differs from Ra-226 to Po-210. In the case of Ra-226, the highest doses are committed by the youngsters aged between 16 and 19 years, while for Po-210 the most

Table 11 Annual effective doses due to the intake Ra-226 and Po-210 for subjects of different ages living in Valea Calugareasca area compared to those living in a control area

Crt. no.	Group composition	Age group	Effective dose (µSv/year)			
			Ra-226		Po-210	
			Plant area	Control	Plant area	Control
1.	Children	1-3	11.4	5.9	61.5	35.1
2.		4-6	10	4.7	37.6	21.1
3.		7-9	10.6	5.3	42.1	23.6
4.		10-12	17.2	8.3	30.9	16.8
5.	Youngsters	13-15	19.4	9.2	34.8	18.6
6.		16-19	37.1	17.7	21.5	11.8

| 7. | Adults | | 8.6 | 4.0 | 19.9 | 10.5 |

exposed group are the children from 1 to 3 years old. In both cases, the lower doses are committed by the adults, which is to be expected due to the period taken into consideration for the dose commitment. In both cases (for children and adults), the doses from the area of interest are almost double the doses from the control area.

To highlight the contamination in the area and the transfer from the environment to the humans, we have also prelevated human embryos and foetuses aborted and determined their radioactive content in correlation with the intake of the mother (see tables 12 and 13).

As we have shown in some previous papers, we could not find any mathematical correlation between the content of the embryos and the age of the mother or the age of the embryo. This may be due to the poor statistics implied by the relative small number of samples.

Table 12 Ra-226 content in human embryos from Valea Calugareasca area

Crt. no.	Mother age	Ra-226 intake (mBq/day)	Age of the embryo (weeks)	Ra-226 in embryo (mBq/g tissue)
1.	24	80.6	5	0.41
2.	25	80.6	6	0.74
3.	24	80.6	6	1.22
4.	24	80.6	6	0.33
5.	22	80.6	6	1.45
6.	26	80.6	6	0.89
7.	32	80.6	6	1.63
8.	25	80.6	6	2.12
9.	25	80.6	7	0.48
10.	40	80.6	7	1.75
11.	31	80.6	7	1.49
12.	28	80.6	7	2.34
13.	25	80.6	7	0.29
14.	22	80.6	7	0.95

Table 13 Po-210 content in human embryos from Valea Calugareasca area

Crt. no.	Mother age	Po-210 intake (mBq/day)	Age of the embryo (weeks)	Po-210 in embryo (mBq/g tissue)
1.	23	40.2	5	0.15
2.	19	40.2	5	0.41
3.	41	40.2	5	0.22
4.	25	40.2	6	0.18
5.	25	40.2	6	0.25
6.	33	40.2	6	0.31
7.	37	40.2	6	0.52
8.	22	40.2	6	0.11
9.	27	40.2	7	0.18
10.	25	40.2	7	0.34
11.	28	40.2	7	0.16

However, when we compare these values to the values obtained for the embryos prelevated from mothers living in an uranium mining area, it can be seen that the values are sensibly comparable, both for the content in Ra-226 and for the content in Po-210. The uranium mining area studied previously by us was polluted with waste from the uranium mines, so we can conclude that the waste from the phosphatic fertilisers plant gives rise to a pollution comparable to the one from the uranium mining area.

4. CONCLUSIONS

From the study above we can conclude the follwing:

- the area near ROMFOSFOCHIM is subjected to radioactive pollution;
- the source of pollution is now given by the phosphogypsum from the unprotected waste repositories, which due to the meteorological factors is spread in the whole area, continuously;
- the amount of Ra-226 and Po-210 of the environmental factors from the area studied is 2÷3 times higher than for those in the control area;
- the amount of Ra-226 and Po-210 ingested by the population from the area of interest is roughly two times the amount ingested by the population from the control area;
- the ingestion of radionuclides results in higher internal doses for all age groups in the subject area than for the people living in the control area;
- for all age groups the highest doses are due to Po-210, although the highest ingestion was for Ra-226;
- the children are the most exposed at the internal irradiation, the highest doses being for the 1÷3 years age group for Po-210 (62 µSv/year) and for the 16÷19 years age group for Ra-226 (37 µSv/year), while the adults have the lowest doses both fore Ra-226 and Po-210;
- the amount of Ra-226 and Po-210 in the embryos is comparable to the amount in the embryos prelevated from an uranium mining area, which is another indicator of the fact that the area near ROMFOSFOCHIM has a higher radioactive level than a normal area.

After identifying the source of pollution and the risk for the health of the population (especially for the children) from the area of ROMFOSFOCHIM Valea Calugareasca, the next necessary step is to continue the study to identify the best solution for mitigating the effects of the pollution and for neutralising the source. The best method should have *a sound ballance cost/benefit.*

5. REFFERENCES

1)   *** E.M.L. Procedures Manual (1990); Ra-02   Radium -226 , Chromate Method 4.5 128 - 4.5. 135.

2)   *** E.M.L. Procedures Manual (1990); Ra-03   Radium -226, Emanation Procedure , 4.5 135 - 4.5 143.

3)   *** E.M.L. Procedures Manual (1990); Ra-07 Radium -226 in urine and water ; 4.5 162 - 4.5 171

4)   *** E.M.L. Procedures Manual (1990);   Po-02 Polonium -210 in water , vegetation Soil and Air filters; 4.5 81 - 4.5 84.

5)   *** E.M.L. Procedures Manual (1990); Pb-01 Lead 210 in Bone, Food. Urine . Feces , Blood , Air and Water. 4.5 73 - 4.5 81.

6)   International Commission on Radiological Protection;   Limits for intakes of radionuclides by workers . Oxford; Pergamon Press ICRP Publication 30; 1978

7)   International Commission on Radiological Protection. Age-dependent dose to member of the public from intake of radionuclides; Part 1 ingestion dose coefficients. Oxford; Pergamon Press ICRP Publication   56; 1991.

8)   International Commission on Radiological Protection. Age-dependent dose to member of the public from intake of radionuclides; Part 2   Ingestion dose coefficients. Oxford; Pergamon Press ICRP Publication   67; 1993.

9)   McDonald , P.; Fowler,S. ;Hyraud, M. ; Raxter M.S. Polonium-210 in mussels and its implication for environmental alpha-autoradiography. J. Environ. Radioactivity. 3. 293-303 ; 1986

10)   SR 13327 . Air quality. Fallout. Determination of polonium 210 content.

11)   SR 10447-3 . Water. Determination of radium 226 contents.

12)   Toader M; Vasilache R ; Considerations on the transfer of natural and artificial radionuclides from mother to embryo - Rom. J. Biophysics 1998 (in press).

13)   United Nations Scientific Committe on the Effect of Atomic Radiation . Sources, effects and risk of ionizing radiation. New York United Nations Publication : E 94 .ix .2 1988 56-78.

14)   United Nations Scientific Committe on the Effect of Atomic Radiation . Sources, effects and risk of ionizing radiation. New York United Nations Publication : 1996, 62-87